God's Wonders

Written and
Illustrated by
Christine Bojahra

The Monarch

Dedication

This book is dedicated to the glory of God

and to my sister Rose

who gave me all the help and encouragement.

God's Wonders!

Volume One

The life cycle of a

Monarch Butterfly

Caterpillar,

I see you crawl.

Why are you on

this garden wall?

You're not afraid of me at all,

Even though I'm big

and you're so small.

You're very careful not to fall,

But you really shouldn't

be there at all.

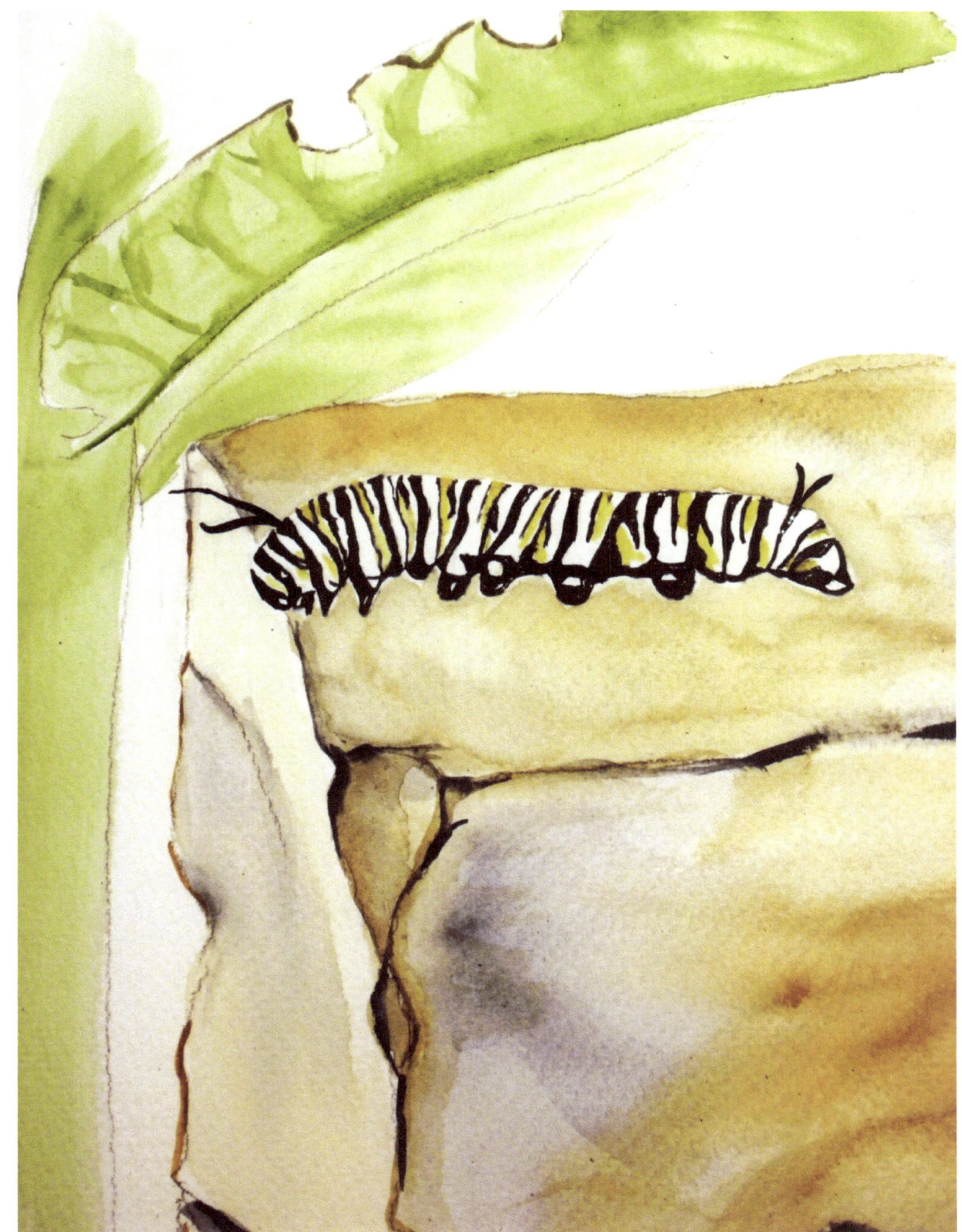

Here is a plant;

I can see where you bite,

This is the place for

you alright.

I'll gently place you on

my hand,

And on this leaf you are

going to land.

Now you are there,

you can eat that.

Eat until you're very fat.

When you're fat enough,

you'll shed your skin,

Your change to a chrysalis

will begin.

Then in your chrysalis

you will hide,

And change very slowly

while inside.

God knows just how

you'll change within,

And soon your new life

will begin.

When you come out,

why you will see,

A beautiful butterfly

you will be.

Then up into the air,

you'll go…

on a very long trip

to Mexico.

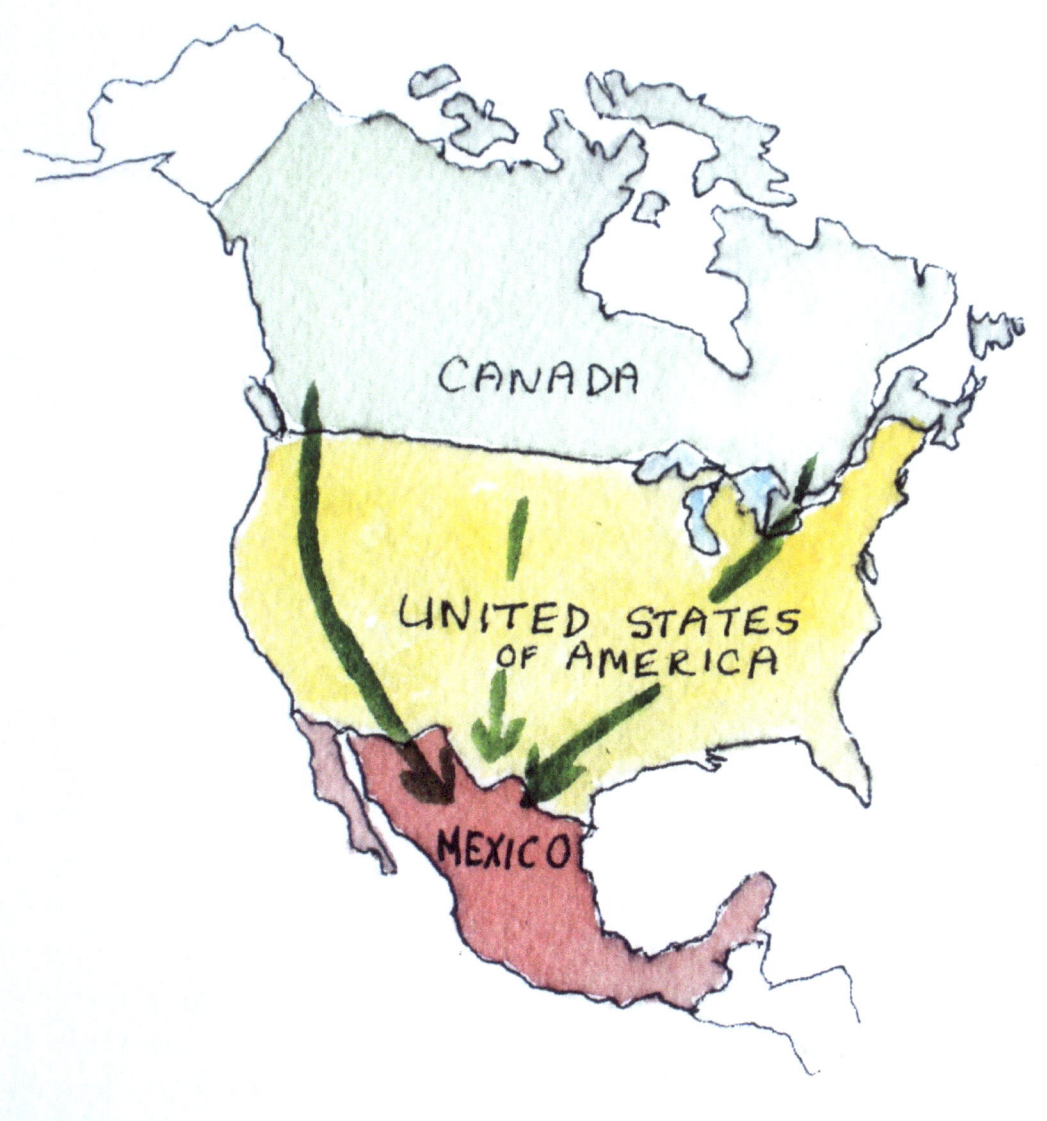

MIGRATION ROUTES

One of the wonders

of God's earth!

The miracle of

a Monarch's birth!

The Monarch

The mother Monarch lays her eggs on the underside of the leaf of a milkweed plant. These eggs

egg

hatch in about three days.

They hatch into caterpillars.
For about two weeks the
caterpillars eat the leaves of the
milkweed plant. Poison in the
milkweed plants make the
caterpillars poisonous
so that birds will not eat them.

This poison also gives them their bright colours, especially the bright orange in the butterfly.

After about two weeks of eating, and it has grown very fat,

It hangs upside down,

and then sheds its skin.

Out comes its chrysalis.

This happens very quickly.

Inside the chrysalis the Monarch begins to change so that it slowly becomes a butterfly instead of a caterpillar.

This takes about two weeks.

As the caterpillar changes into a butterfly, the outside skin of the chrysalis becomes very thin so that you can actually see the butterfly inside. This skin then breaks open and the butterfly emerges.

At first it is all rolled up and very wet. It takes a long time before the butterfly can spread its beautiful wings.

As the wings dry, the heart

pumps blood into them to give

them strength.

Soon the wings spread open

and the new butterfly

is now ready to fly.

It feeds on flowers for about six weeks and then lays its eggs on the milkweed plant again to create a new generation.

Each summer there are usually four generations of butterflies.

The fourth generation, born in September or October does not lay eggs but flies to a warmer place, usually Mexico.

This is called migration.

The Monarch stays there all winter and in the spring it flies back north and begins the cycle all over again.

One of the wonders of God's

earth! The miracle of a

Monarch's birth!

God has many wonders on this earth. The Monarch butterfly is only one of them.

The life cycle of the Monarch reminds us of the re-birth that Jesus talked about.

"I assure you, unless you are born again, you can never see the kingdom of God."

John 3:3 (NLT)